我知道，我爱

总有一天你会俘获我的心

透过你天窗上的星辰

温和的灯光洒落在古典的

梳妆镜前，如你的凝视深入到我的梦境

自落地窗前、纯白色的纱帘下

透进来的月光是你的信使

带来了你

埋藏已久的心事

我挨着闪耀如银河的墙上

如沐浴在灿烂的爱里

我沉思着

眼中充盈着喜悦的泪

这都是你在向我述说的爱情

摘取并改编自泰戈尔的《渡口》

A touch of Anita Tang

写给空间的情书

贝思出版有限公司 汇编 侯阳 赵伟韬 主编 赵耀 金品卓 译

辽宁科学技术出版社

Even today, the design industry of Hong Kong seems to be dominated by men between the ages of 40 and 50. It is challenging for a woman to cut a niche in the professional market, but Anita Tang is a rare exception. Her indulgence and originality has produced unique designs that carry an unspoken sophistication of elegance and extravagance. Her first collection of design projects "A touch of Anita Tang" focuses on residential designs. It contains over 10 household and model home projects which not only expresses the talents of a female designer, but also the prevailing grace and passions exclusively found in feminine practitioners.

香港的室内设计行业至今仍是由40~50岁的男性主导，女性要在这行业突围而出殊不容易。但是邓凤贤是一个罕有的例子。“用心、创新”是她的设计原则，出自她手笔的设计总是独一无二，同时散发出一种说不出来的高贵华丽。《写给空间的情书》是她首本个人设计专集，是一本以“家”为主题的空间设计书，包含了十多个家居、家居示范单位设计案例，当中不但呈现出一个女性设计师的才华，亦展示了只有女性设计师才会流露的优雅和对空间的情感。

CONTENTS 目录

Preface

Anita Tang Fung Yin

She possesses an adorable profile, smart mentality, and a name of potent significance. With long curly hair which resembles the gentle seaweeds, just a casual outfit of white shirt and black pants is already a statement of artistic indulgence. With a dedicated passion for aestheticism, she decided to study in Canada majoring in Interior Design in early adulthood. Upon return to Hong Kong, she established her own design company and tapped into the Mainland design market. Currently the Deputy General Manager of Sales Department in Henderson Group, she oversees the design and aesthetic components of building complexes. Self confident, yet articulated in social etiquette, she maintains impeccable interpersonal networks. Although she values good tastes and quality of life, and enjoys spending time in art galleries, tasting red wines, and appreciating branded antique watches during spare time, she is passionate in the care and well being for under-privileged people living in endangering environments.

The intention of publishing the first personal design collection is to convey some messages. She hopes to record the stages of her design in a summary and to share her thoughts, experiences, and accomplishments with professional counterparts, working partners, as well as friends and relatives. It is consistent with her theme in the earlier column writings for Ming Pao; the only difference is that the last attempt was a narration of laughter and tears in design projects, while this time, with the assistance from Pace Publishing, she could visualize her values and aestheticism in spatial treatment through graphics and visual stimulations to trigger interactions and inspirations.

It is not an adaptation for reality, but the persistence on design hypothesis. Looking back on her design career, from commercial designs to residential and model home designs, it may seem to be a step backward for others who perceive it as a narrow path from the main stream. For her, the design of residential units and model homes has provided tremendous autonomy and job satisfaction to actualize methodologies and ideologies that cannot otherwise be accomplished in a commercial setting. Furthermore, these projects link her directly with the community, the group of people who utilizes her design. The contact is direct and extensive ... for a designer, nothing is more important than that.

From office design to model home design, during these 20 years of professional dedication, some subtlety is delivered to her, which was expressed in a former interview; it makes her feel the importance and contribution of her work. She has a mission on residential home design: to create an appealing and comfortable living environment for people. Almost all of her projects are infiltrated with a touch of grace, and the sensations of an elegant lifestyle due to her total indulgence of passion into the spatial design; and the end product is thus turned into a gem of obsessions.

With utmost sincerity,"A touch of Anita Tang" is dedicated to you all.

Michelle Ng
Editor, Pace Publishing Ltd.

前言

邓凤贤

她拥有一个气势十足的名字，也拥有美丽的轮廓和聪明的头脑。一头海藻似的长曲发，随便穿一套白衬衫黑长裤就能突显其艺术气质。对美有要求的她，少女时期已决定赴加拿大主修室内设计。回港后自组设计公司，也开展在内地的设计。现任恒基营业部副总经理，专门从事监察楼宇的设计与美学事项。她充满自信，懂得人情世故，面面俱到。闲时爱逛画廊，品赏红酒，鉴赏古董名表。她重视生活品位和素质，也关顾活在水深火热的人们的福祉……

写给 –
出版首本个人的设计师专集，当然想借它传达一些信息。她最想借这本书，为自己的设计阶段做一个小结；与同业好友、工作伙伴、知己亲朋，一起分享她在设计路上的想法、经历和成果。这和她早前在明报副刊专栏写作时怀着的想法是相似的，只是上回借由文字和盘托出设计工作中的喜乐辛酸，今次则希望以视觉画面来表现她对空间的观念和美感，借此与这个行业的有心人生出同感和互动。

空间的 –
不是一种对现实的妥协，却是对设计的执着。回顾过往的设计路，由商业空间设计走向家居和家居示范单位的过程，在一般人眼中，可能是往回走，从一条宽阔的路走进一条窄路。然而在她心里，设计家居示范单位却给予她极大的满足感和自由。因为，她不但可以将那些在商业空间里未能实践的设计理念和构思，在家居示范单位中实现，而且，她的作品让她接触到群众——最终使用她的设计的人，是直接的、完全的，作为一个设计师，最重要的莫过于此。

情书。
20年的设计经历，由以往主打的办公室设计到今天的家居示范单位，对她而言有一种更深层的意义，她曾在一个访问中提及，感到自己所做的事情有贡献是重要的。设计家居让她怀着一份使命，就是希望为人们缔造一个舒适和快乐的生活环境。也几乎每一个空间，从她的设计图纸里实现出来的，总渗透着一种说不出来的高贵，一种优雅从容的生活质感。这是因为每一次她总毫无保留地为空间设计投递了最真挚的情感，成就了一个个感情结晶。

现诚意将邓凤贤《写给空间的情书》献给您。

贝思出版有限公司编辑部　吴嘉慧

Diary 日记

邓凤贤的工作繁忙，每天也面对着很大的压力，然而她别有一套减压方法，就是每天冥想半个小时，停下来让自己平衡心神，在令想法清晰之余，也会把事情做得更好。

Overwhelmed with work and related pressures, Anita has a unique way of pressure reduction. She stops her activities and meditates for half an hour to maintain an internal balance. Besides clearing her mind, it helps to make

15:00 比华利山别墅1街81屋 示范单位装修工程议价会 (Rm 7305, IFC II)

她曾为多个大型豪宅设计示范单位，包括比华利山别墅、麦当奴道11号及嘉亨湾等，近日她凭借荔枝角道33号百汇轩示范单位再一次获得大中华杰出设计奖。

She has designed many model homes for luxurious apartments including Beverly Hill, 11 MacDonnell Road and Grand Promenade. Her recent project, Cite 33 at Lai Chi Kok Road, once again, brought her an award in the "Outstanding Greater China Design".

15:30 The Spectacle PC (新都城中心二期1楼)

她最爱饮红酒，细细品尝美酒的时候亦悟出一些对人生的看法，她认为红酒是充满个性的，出产自同一个木桶的红酒，却可以每一瓶都有不同味道，就像有生命似的。

Red wine is her favourite; wine tasting has always triggered fresh interpretations of life in her. She sees the individuality of red wines. Although brewed from the same barrel, each bottle carries a unique taste as if it possesses a life of its own.

香港的事业女性不易为，一方面既要在工作上与男士“平等”，为事业努力拚搏，另一方面又要满足家人的期望做贤妻良母。

It is a challenge to be a career woman in Hong Kong. On top of the fierce competition with men in the career market, she has to satisfy the expectations at home to be a qualified mother and housewife.

Caressed by the re
environment, it ma
hassles and dazzle
life.

Grand Promenade (Outstanding Greater China Design Award 2007)

With a panoramic sea view, the designer applies a symmetrical minimalist design integrating nature with urbanity to promote contemporary grandness as well as sophisticated elegance.

White in color with dark brown accentuations, the show flat displays vertical linear graphics as the design theme combining mirror and crystal reflections to boost visual excitements. The designer has painstaking experiences in producing unique motifs through the use of diversified materials. The utilization of quality marble, crystal, clear mirrors and synthetic leather wall paper to match the Italian furniture and French paintings are details which reflect her refined precision.

Taking advantage of the symmetrical floor plan, the designer adopted a symmetrical model with minor modifications. Both sides of the lounge area are furnished with identical sets of sofa, area rugs and crystal chandeliers. Although one side leads to the VIP room while the other connects to the conference room, they are presented with symmetrical identity. The curvilinear corners on both sides are allocated as rest areas for people to enjoy the sea views.

Located at the centre is the sales counter which faces the balcony. A translucent Corian panel that displays an elegant embossed graphic is erected behind the counter. To carry through the symmetrical design, the left side of the counter is an open kitchen with a bar area decorated with suspended crystal curtains. On the right is a glowing glass staircase that connects the interior to the terrace garden. The wall affixed to the staircase is covered by alternating stripes of clear and tinted grey mirrors to generate a contrast, and at the same time, reduce the intensity of glare. To enrich the sensation of extravagance, a grand chandelier that is suspensed from the ceiling to the entrance of the staircase on the lower level is installed. The terrace garden is defined into a dining area and seating area complemented with a Jacuzzi and some rattan furniture. Caressed by the relaxing environment, it makes you forget the hassles and dazzles of cosmopolitan life.

laxing
kes you forget the
s of cosmopolitan

嘉亨湾 （荣获2007大中华杰出设计大奖——室内设计）

这个住宅拥有广阔的海景，设计师采用简约对称的设计概念，融入自然和现代都市的豪华摩登感觉于一室，且不失瑰丽高雅的气派。

这个被用作销售楼盘和展示设计的单位以白色为主，深咖啡色作为衬托。纵向的线条图案是整个设计的重点，设计师借物料反映出来的镜光与水晶折射增加了视觉上的趣味性。她一向擅于利用物料的可塑性来打造与众不同的空间，高质的云石、水晶、清镜，以及仿皮质墙纸，配合来自意大利的家具和法国的挂画，她在每一个细节上都是一丝不苟的。

据设计师说，原来的图是一个左右对称的空间，所以她想利用这个特点做一个左右对称的设计，只将一小部分的地方作修正。客厅的左方及右方并列两组无论体积大小及设计款式都相同的沙发、地毯以及水晶吊灯。两旁通往贵宾房及会议室的入口，亦刻意营造相对的效果。她在两旁弧形的空间，做了一个休闲的角落，供人观赏海景。

销售柜台是这个空间中心点，与露台相对。柜台后的Corian背板，在透光的效果下展现了典雅的浮雕图案。为承接对称的原则，柜台左边的厨房亦改变成开放式，同时加入酒吧区，以垂挂的水晶帘子作装饰。右边的一道发光楼梯，采用玻璃构造，由室内贯通至天台花园。楼梯旁的墙身铺贴了直条子的清镜，不过全部用上清镜会做成映射过剩的效果，同时，为减低其亮度，她用上灰镜与清镜相间，这亦产生了一个对比。为加强华丽感，在楼梯旁挂了一盏长长的水晶吊灯，由上层天花垂挂至楼梯的入口位置。天台花园的布局，分别有进餐区和起坐间，还有一个按摩浴池、一些藤器家具。置身于如此悠闲之地，令人把都市的繁华喧嚣都抛至脑后了。

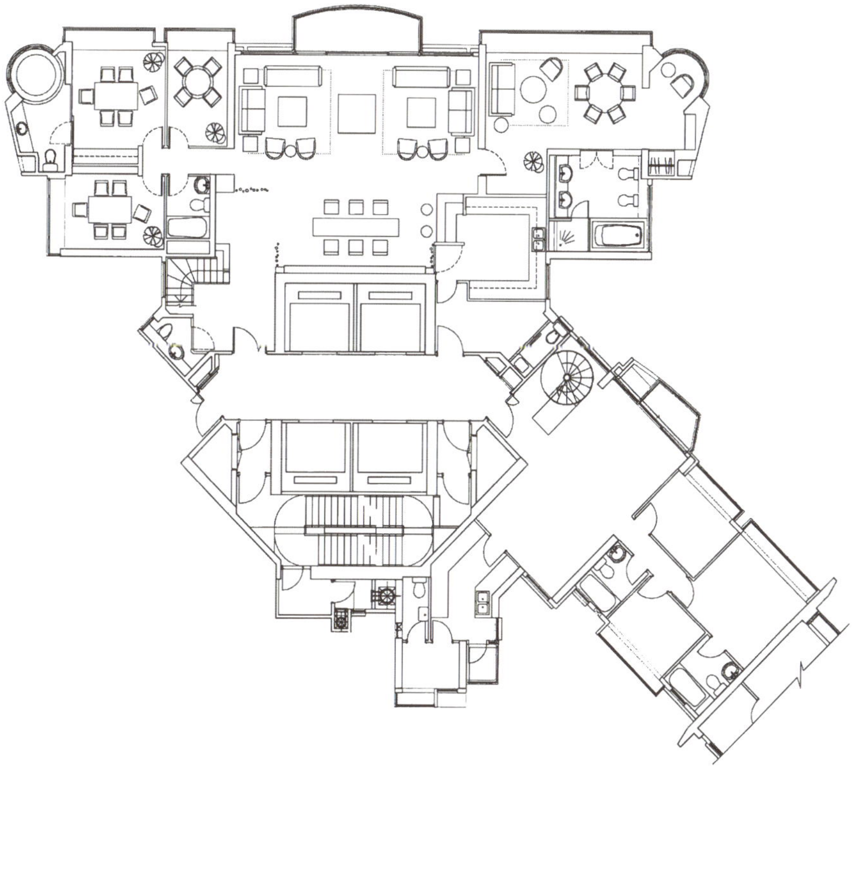

Featuring curvili

feminine design

highly popular w

Residence A at Kings Park Hill

Themed on contemporary extravagance, the designer delivers a sophisticated lifestyle of elegance, romance, and arrogance in this King's Park Hill show flat through the mix and match of black and white. Featuring curvilinear contours, its feminine design is inspired by the highly popular works of Chanel. Material selection include the glamorous and dainty white crystal, elegant black and white marble tiles with snowy graphics, mysterious black walnut wood, and white upholstery that carries imperial royalty.

Possessing an equilateral layout with clearly defined zones, the door entrance is placed in the middle with the living and dining rooms on the side while the living room and hallway will lead to the master bedroom and guest rooms at the back. A closer look at the hall will uncover the designer's attention to the refinement of details. To stimulate a charming appeal, glossy white and brown marble floor tiles are orderly displayed in the living room where a white leather sofa is furnished. Boldly contrasting with the white sofa is the black dining table and chairs in the dining room. Arched in shape, the dining chairs accentuate strong vintage flavor to echo with the carpet in the living room which carries coordinating prints in minimalist approach.

The decoration design of every spot in this space is consistence with the main theme. A dazzling and opulent effect is generated through the combination of grand crystal chandelier, clear mirrors and crystal blinds suspended from the ceiling. With the feature of mirror reflections, crystal blinds on the dining area resembles a glittery waterfall enriching an excited visual effect. The mirror on the ceiling also produces height illusion. Every piece of furniture and fixture in this show flat is properly selected which well demonstrates the painstaking effort of the designer.

In the master bedroom, inspired by the interior design of Chanel boutique, black resin wardrobe with transparent design is engraved with floral prints allowing the penetration of light to promote a sentiment of romance. The beddings and wall coverings are in soft round pattern which dramatize the femininity of the home.

...near contours, its ... is inspired by the ... orks of Chanel.

京士柏山（1）

这个示范单位位于京士柏山，设计主题是摩登华丽，设计师想以黑白制造高雅、浪漫、矜贵的生活质感。这个带有女性感觉的设计，主要以弧线来塑造，设计灵感有部分来自对女性来说总是趋之若鹜的Chanel。设计师在材料运用上亦选取了能营造华丽矜贵的白水晶、高雅的白底黑花纹雅士图白云石地砖、带有神秘感的黑色索色胡桃木皮和充满贵族气质的白色扪皮等。

整体空间方正、布局分明，客餐两厅各据左右，大门在两厅中间。越过厅区和进口走道，便是主卧室与客房的设计。回到大厅，细看空间每个角落，你会发现设计师非常重视细节的处理。为了增加趣味性，亮丽的白云石地砖与浅咖啡色云石地砖在厅堂的地面错落有序地铺设，而客厅位置放一张白色真皮沙发，与餐厅一套黑色餐桌椅组合，恰形成一个强烈的对比。餐椅以弧形线条打造而成，极富古典美，回应了客厅一张别具简约味、花纹相仿的地毯。

空间里不同的角落，也是与设计主题一脉相承的：置于客厅天花中央的水晶吊灯带来的豪华感，以及在餐厅的清镜围身前方、自天花流泻而下的珠帘，其闪烁耀眼的效果为一室的华丽气派写下了最佳注脚。因为清镜的双重反射，为水晶珠帘带来更丰富的视觉效果，使置身餐厅的人们仿若在瀑布流泻的环境下用膳。餐厅的天花铺设了清镜，使楼层看来更高。这里几乎每一个部分都是精雕细琢的，而每一件家具都是设计师亲自为这个示范单位设计的，每一件饰品摆设也是如此的恰如其分。

在主卧室里，衣橱门面上的雕花图纹、从背后透光的设计，为寝区营造了摩登的浪漫气氛；其质料采用了黑色聚脂打造，光亮的表面参照了Chanel时装店高雅的气质。这里的墙布及床铺的圆形柔美线条亦突显了女性的温柔气质。

redwine
whitewine

CHANEL
N°5
CHANEL
N°5
CHANEL
PARIS

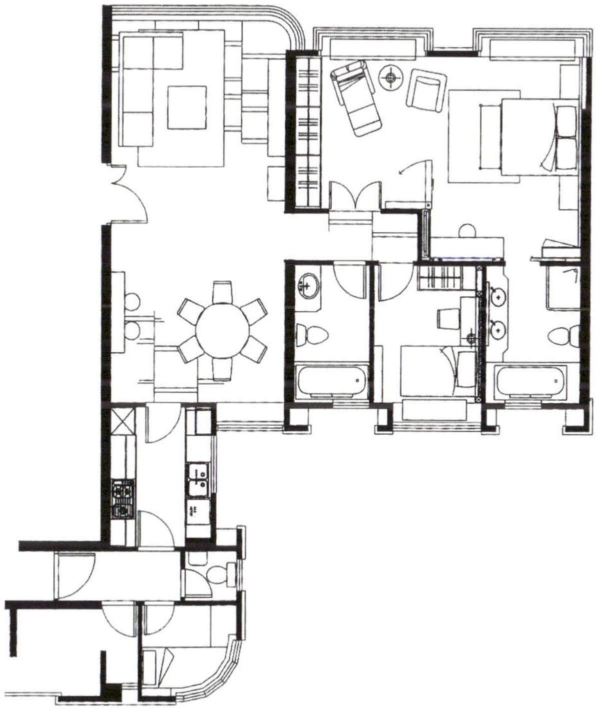

a tasteful and co

themed on moderni

design

Splendid Place

The show flat located at Taikoo is specially catered for the young and single professional or a youthful couple to offer a tasteful and comfy dwelling themed on modernity and minimalist design. A symmetric space is established by combining two adjoining units with one of the kitchens removed to accommodate an expanded master bedroom. Featured in the living room are wine cabinets, extravagant crystal light fixtures, and elegant Italian furniture which blend in with the sauna and Jacuzzi in the master bedroom to promote quality lifestyle.

White, grey, and silver are the main color scheme; complemented by gentle illumination through light troughs and crystal pendent lights, the home delivers a prevailing touch of elegance. The integration of light oak wood and tinted mirror as wall coverings nurtures the furry sofa cushions of black, grey, and light brown for an enriched subtlety. To enhance relaxing sensations and enjoyable lifestyle, the designer specially designed a florescent-lighted wine cabinet for the homeowners.

Quality and painstaking details are found in every corner of the house. In the living room, a decorative fireplace is furnished between two floor-to-ceiling windows; facing the fireplace is a sofa set with adjustable back to watch TV or enjoy the external scenery. Grey tinted mirror is installed on the right of the entrance wall which not only enhances spatial generosity, but also conceals the door to the master bedroom eliminating direct access from the dining room to the master bedroom. The corridor on the left of the dining room leads to a guest room which is separated with the study room by a wooden sliding door to provide a tranquil reading spot for the guest. When asked about her favorite pick, the designer concluded that the translucent glass cabinets outside the master bedroom featuring back-lit light pillars is her favorite as it delivers vibrant and trendy feel.

mfy dwelling

ty and minimalist

汇豪峰

这个位于太古的示范单位看来简洁又摩登，设计师说是专为事业有成的单身一族或年轻夫妇而设，为他们提供一个极富品位而又舒适的居所。她将两个相连的单位打通，成了一个左右对称的空间，又把当中一个厨房拆去，以节省空间来扩大主卧室。厅区的酒柜、华丽的水晶灯饰、高雅的意大利家具，以及主人浴室内的桑拿、水疗设备，全用来为户主塑造高品质的生活。

整间屋的色调以白、灰、银为主，配合散发自灯槽和水晶吊灯的柔和灯光，令这个家弥漫了一种高雅的格调。设计师主要以浅色橡木和灰镜作铺缀墙身的材料，附以沙发上的黑、灰和浅咖啡的毛毛垫子，来丰富生活的质感。为了让户主尽情享受生活，设计师特别设计了一个萤光酒柜，带来了一种轻松的气氛。

贴心的细节散布于家中的每个角落，一具装饰火炉设于客厅里两面落地窗之间，与之相对的沙发靠背可作调较，使人们可按自己的喜好看窗外景色或观赏电视节目。设计师在大门右方的墙身铺贴了灰镜，一方面加强了空间感，同时也隐藏了一道通往主卧室的暗门，避免了由餐厅直达睡房的突兀感。从餐厅经过左方的走廊，会到达客房，它和旁边的书房以一道木敞门阻隔，为客人制造了一个宁静的阅读环境。问及设计师一个喜爱的细节，她说是主人浴室门前两个通透的玻璃装饰柜，它背后的光柱营造了时尚的感觉。

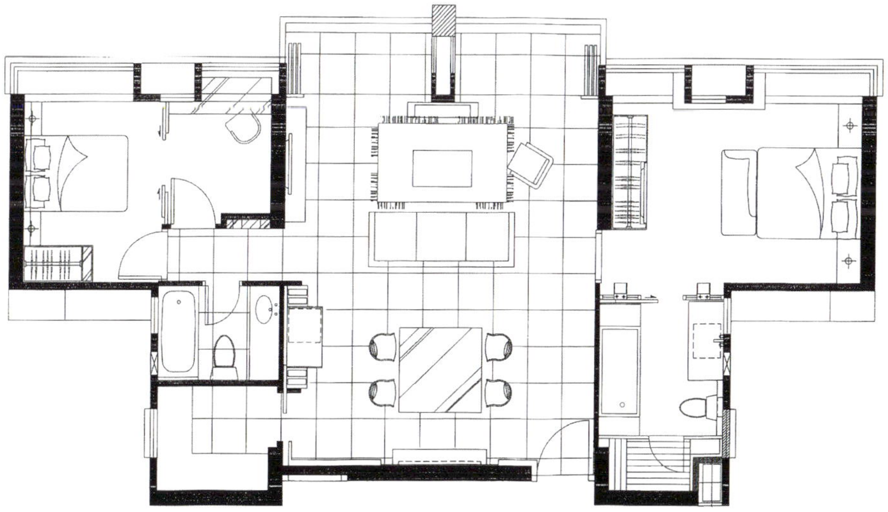

a glamorous touch complemented with of the Renaissance

Residence A at MacDonnell Road

This show flat unit offers a glamorous touch of gold complemented with an artistic taste of the Renaissance era. The designer has successfully made an innovative attempt to meet the client's needs, and at the same time provide a new definition to the conventional perception of outdate and intrusiveness in gold. Infiltrated with design elements of the Renaissance and Louis XIV, its artistic flair submerges the arrogance and conceit of glittering gold and turns the home into a European art gallery.

A classical golden mirror is placed at the wall of the foyer to echo with a pair of crystal chandeliers which generates a warm welcoming touch. In the living room, a three dimensional wall in classical palatial design complements the white fire place and white sculptural column to accentuate the home with modish elegance that supplements the glamour of gold.

Replicas of Renaissance sculptures and paintings are explicitly displayed in the living and dining areas; and the display and installation of which has consumed painstaking efforts from the designer. The painting in the living room, for instance, is suspended behind the sofa backed by the draperies generating an illusion of being floating in the air. A row of sculptures placed at the back of the dining chair is dramatized by the spot light to trigger an appealing artistic ambience.

Wallpaper with gold leaf prints coordinates well with the rectangular light trough and extravagant rose shaped crystal chandelier to deliver an elegant statement of lavishness.

Although the marble fireplace in the living room is installed for decorative purpose, the two square pillars with gold leaf prints that extend from the floor to the ceiling trigger a grandeur palatial mood. All furniture selected is classic European designs to maintain a consistent theme. The master bedroom, which upholds the design concept, features draperies that carry rich characteristics of vintage European flavor while the study room promotes the sentiments of relax and leisure.

l of gold

h an artistic taste

ce era

麦当奴道住宅（1）

这个示范单位予人以金碧辉煌的感觉，且充满文艺复兴时期的艺术色彩。它是设计师一个突破性设计：既迎合了客人的需要，同时又跳出了金色所带来的俗气、老气的框框。在这个空间，她注入了路易十四及文艺复兴时代的设计，用艺术元素中和了随着金碧辉煌而来的炫耀与浮夸，从而营造一个仿如欧洲艺术馆的家。

在大门的玄关位置，设计师挂了一道古典的金色镜框和一对水晶吊灯，这增加了玄关的空间感，并产生一种迎客的感觉。来到大厅，这里的墙面是仿古典宫廷的立体墙身，与白色的壁炉及雕塑下的白色立柱，俱增加了室内的典雅气派，更与耀眼的金色互相辉映。

据设计师说，客厅与餐厅的最大特色，是置放了不少仿文艺复兴时期的雕塑和名画，在摆设的过程里也花了不少心思。就好像客厅墙上的一幅画作，她先选用了窗帘作为背景，再凌空垂吊在沙发背后，并没有贴墙而挂，使这幅画看起来有浮游在半空的感觉。置放在餐椅背后的一列雕塑，被天花的射灯映照着，灯光则聚焦在雕像上面，使这里弥漫着一种艺术氛围。客厅、餐厅的天花铺贴了金箔墙纸，其方形灯槽配合高贵夺目的玫瑰花水晶吊灯，令空间华丽而不庸俗。

虽然客厅的云石火炉只作装饰之用，但两条贴上金箔的方柱，由地台伸延至天花，带来了宫廷式气派。全屋家具都是仿照欧陆古典家具的样式设计，呼应主题。走进主卧室，窗帘的设计带有浓烈的欧洲古典特色，而书房的布置以营造悠闲的感觉为主。

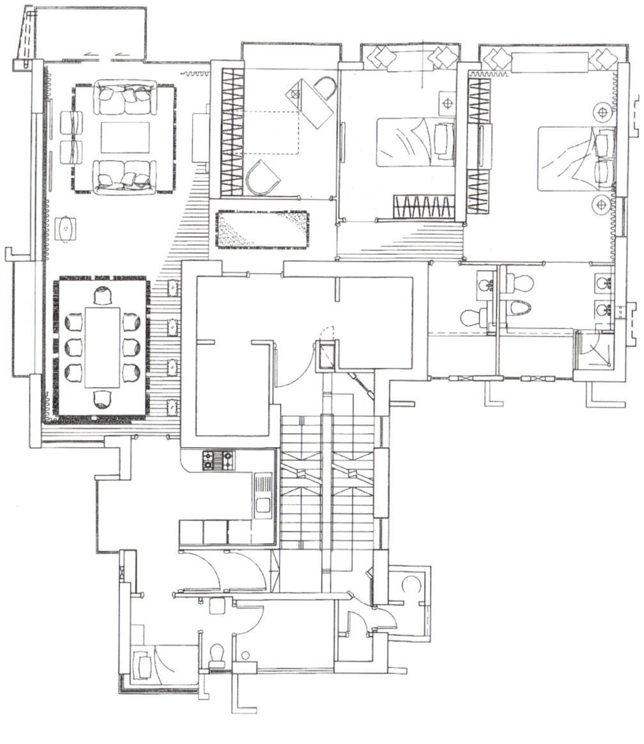

Taking advantag
textural sensation
color scheme, the
promote an appea

Villa Costa

White is the client's favorite color, therefore he made a seemingly simple request to the designer that white be applied throughout his 3-level home ranging from interior décor to furniture and fixtures. The challenge to the designer lies in its easy snare to being overly cold or association to a funeral home. Taking advantage of the varying textural sensations even on the same color scheme, the designer is able to promote an appealing mood in the monochrome tone.

A good illustration can be found in the living room where glossy white marble flooring is complemented with beige wall covering of matt wood. Further enhanced by supple fabric tenders, a bold contrast of textural disparity is triggered while the clear mirrors provoke an interesting overlay of spatial configurations.

Furniture selection includes a coarse white leather sofa set and side table made of canvas which suggests a sharp textural distinction to the white fluffy carpet. To bring out a sense of splendor and warmth, off white ceiling troughs emitting yellowish illumination are installed to complement the deluxe chandelier for an elegant statement of extravagance. Last but not the least, covered by Thai silk valance and panels in light embedded prints, the high windows are highlights of the residence.

e of the varying

s even on the same

designer is able to

ling mood

蔚海山庄

客人喜欢白色，他向设计师提出了一个看似简单的要求：在这幢三层高的独立别墅里，其装潢和家具想单一地用上白色。当时设计师的确感到有点苦恼，因为若处理不当会令整个家变得冰冷。最后设计师想到的是采用同色系却不同物料的质感，在淡浅的色调上，创造出富有趣味性的环境。

就以客厅为例，先用上光面的白云石作地台，配上亚面浅米白色木纹墙身，再在当中加上柔软质地的布料，不同质感形成强烈对比。另外，墙上铺贴了清镜，利用重叠空间带出空间上的层次感和趣味性。

至于家具方面，则以粗糙的白色真皮作为沙发用料，帆布造的茶几及地上铺上长毛浅白色地毯利用质感营造出强烈的对比。为了带出豪华和温暖的感觉，纯白色的天花装设了带米黄光的灯槽，使灯光柔和舒适，加上名贵的水晶灯，华丽而不俗气。最后，高窗特别加浅暗花泰国丝质窗帘和窗纱，亦是重要的点缀。

PHILIPS

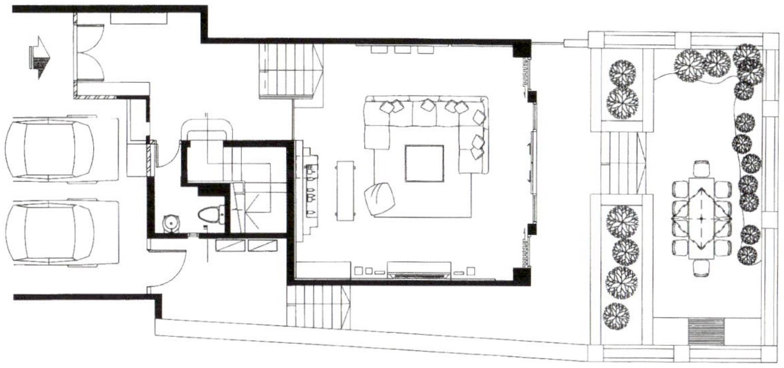

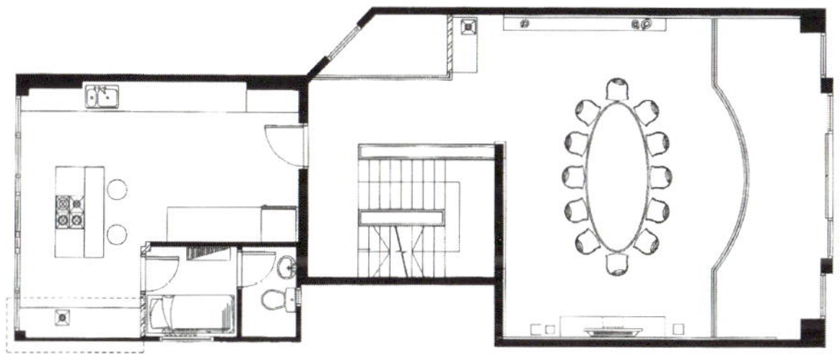

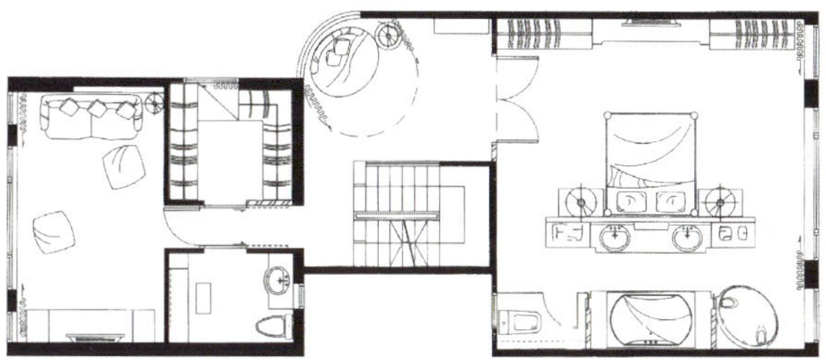

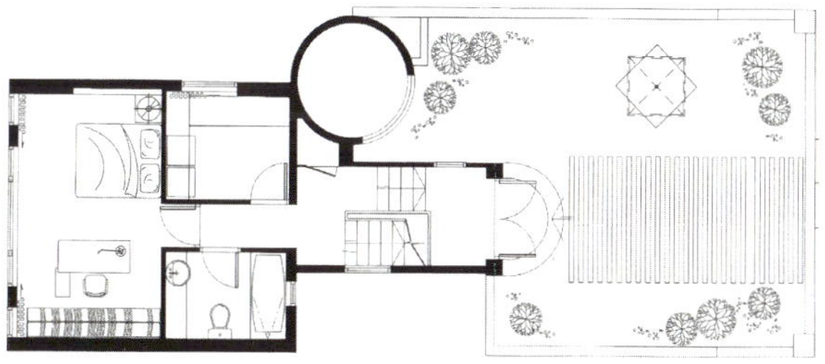

as an artifact of contemporary sophi[s]... and stylish express[ions]

Cité 33 (Outstanding Greater China Design Award 2009)

The designer intended to deliver the sales office as an artifact of sculpture promoting contemporary sophistication, trendy and stylish expressions. Applying a combination of clear glass, tinted black and gray mirrors in alternating vertical strips as wall covering, the crisscross of shades, parade of stripes, and low-rise furniture not only dramatize the generous headroom, but also nurture the sensations of grandness and audacity. Complemented with jade like translucent marble floor tiles, a touch of exquisiteness is explicitly demonstrated.

An intricately engraved wooden wall panel in white piano lacquer is erected for the reception counter at the entrance of the sales office to promote extravagance. The designer revamped the original layout of 4 rooms into 3 to accommodate two large conference rooms and a small meeting room. Some of the hard walls in the hallway are removed and replaced by glass panels for visual expansion; and to intensify the tiered effect, ceiling height of the hallway is lowered. Glittering fancy wallpapers are used in the conference rooms to express a statement of contemporary upbeat. One of the conference rooms is mounted with an original oil painting in violet to promote stylish embellishment.

In order to allow the sales office to be turned into a future show flat, the three conference rooms can be transformed into bedrooms and study room, the furniture are also portable for easy replacement. A kitchen with full amenities is also included in the layout. Taking advantage of the 4M high ceiling, the designer provides ample storage capability by installing rows of hanging cabinets affixed to the ceiling. To avoid an elongated visual impact, one of the walls is covered by clear mirrors to balance off the visual breadth. The ceiling moulding, which integrates classic intricacy and contemporary sophistication, is extended to highlight the generous headroom.

The infiltration of splendid extravagance is closely associated with the fine selection of furniture and fixtures such as the trendy and elegant chandeliers in the lobby, the white leather executive desk and chair in the conference room, and the striped area rug in the sitting room. They are either hand picked or specially designed to match the overall plan.

sculpture promoting
stication, trendy
ions

百汇轩 （荣获2009大中华杰出设计大奖——室内设计）

设计师希望这楼盘销售中心可被塑造成一个有如雕塑一般的艺术品，主题上表达出现代化、新潮与时尚。所以，她在物料上选用了黑镜、清镜与灰镜，垂直地铺贴墙身；色彩上的交错、条子的形态以及其矮身家具的配置均夸张了这个项目的高度，同时也加强了它的宏伟气度，营造出活泼的效果。地面铺设像通透玉石一般的云石，使这个设计项目更显精致。

销售中心的入门处，设计师在接待的位置塑造了一面细致的墙身：在木雕花墙面上，涂上一层白色钢琴漆，带出高贵的感觉。在空间规划上把本来的四房改装成三房，用途分别为两大会议室和一小会议室。在间隔上把走道部分的实心墙拆了，改装为玻璃间隔，在视觉上看来宽阔一点。同时，为了加深空间的层次，她刻意把这走道的天花压低一点。在会议室内，设计师用的是华丽闪耀的墙纸，因为她想带出一种动感时尚的气氛。其中一间会议室更挂上一幅以紫蓝为主调的油画，是由画家亲自绘画，来配合这个充满风格的空间。

为了方便它随时可被改为住宅的示范单位，三间会议室均可被改装成睡房及书房，这里所有的家具也是活动式，可轻易转换成其他家具。另外，这里有配备完善的厨房，基于本身有达4米高的楼层，设计师也特别做一列贴着天花板的吊柜，来增加收纳空间。为免极高的楼层只得窄长的效果，设计师把一边墙身铺贴了清镜，以图拉阔空间的宽度，亦在视觉上将比例合理化。为突显高楼层的优点，厅区的天花线据设计师所言是刻意加粗的，这样安排也可使古典的精致和当代的时尚风格交相融合。

这个空间所流露的高贵华丽，与当中的家具饰品也有莫大的关系：厅区的天花中一盏既华丽又时尚的水晶吊灯、会议室内的白皮革书桌及椅子、客厅的一幅条子地毯，都是特别为配合整体设计而编织或塑造出来的。

The most eye-cat

is undoubtedly th

accentuated by S

Residence A at Grand Waterfront (Outstanding Greater China Design Award 2008)

Themed on diamonds, the show flat and sales office promotes sensual affection for glitters in the holistic proposition. Naturally the priciest unit due to its top floor location, the designer heightens the lavish splendor to boost its supremacy. The most eye-catching focal design is undoubtedly the featured wall accentuated by Swarovski crystals; its glamorous appeal complements the shimmering wallpapers and silvery mosaic ceramic tiles to deliver a proud statement of extravagance.

Although the show flat and sales office takes up only 400 sq m, it is characterized by numerous unique features as demonstrated by the exceptional interior staircase leading to the terrace. Walled by clear mirrors at the turning corner of the staircase to widen spatial breadth, the opposite wall boldly presents the latest crystal mosaic tiles which contrasts with the naturally toned grey pony hair wallpaper on the upper floor for an enriched design presentation. The application of ceiling mirrors generates reflective images from all directions; and since the home generously features translucent glass and mirrors as prime materials, Fior Di Pesco Carnico marble tiles are chosen for flooring. Their distinguished grains stimulate a sharp disparity to the translucent and glittery components of the interior.

Excitement is found everywhere ranging from decorative fixtures to design elements. The reception area, for instance, displays a modern and stylish crystal chandelier suspended from the ceiling; even the dainty tables sporadically placed at different spots are designed like transparent jewelry counters with featured crystals making home purchase as exclusive as jewelry shopping. The penthouse terrace which features floral selection of strong European flair offers a cozy environment for tranquil relaxation.

tching focal design
e featured wall
warovski crystals

翔龙湾（1）（荣获2008大中华杰出设计大奖——室内设计）

钻石，是这个示范单位销售中心的设计主题，所以整体空间予人的感觉是闪闪生光的。这里位列顶层，自然而然成了最高价单位，所以设计师特别把它做得很贵气耀眼，来突显其身价。当中最触目的是一幅表面以施华洛世奇(Swarovski)水晶铺缀而成的墙身，制造了一个很漂亮夺目的效果，呼应了室内闪亮的墙纸与银色马赛克瓷砖，华丽的气质流泻一室。

其实这个示范及销售中心的空间不大，占地400多平方米，但设计上特色各异，只一条内置通往天台花园的楼梯，已予人以不同的感受。设计师在楼梯的转角处铺贴一列清镜，作为拉阔空间之用。楼梯旁的一面墙上也大胆地使用了最新的水晶马赛克瓷砖铺贴，走到楼上一层却换成了富自然质感的灰色马毛墙纸，这些设计细节令整体效果更为丰富。天花的清镜使用，令四周也呈反射状态。因为整个设计多采用通透的玻璃与镜材物料，所以设计师选取了柏斯高灰云石砖作地面，它的纹理鲜明，与闪亮、透明的质料形成了强烈的对比。

在这个空间里，无论设计装潢或是饰品家具，都叫人感到惊喜。就好像接待的位置，天花悬垂的是一盏既摩登又别具气派的水晶吊灯。至于那些设在不同角落的桌子，设计师也将之设计成用来看珠宝的透明柜台，摆放了闪亮的水晶，予人一种买楼好像买珠宝般的富贵之感。走到天台花园，这里被塑造了一个悠闲写意的环境，种植在这个花园的花草充满了欧式风格。

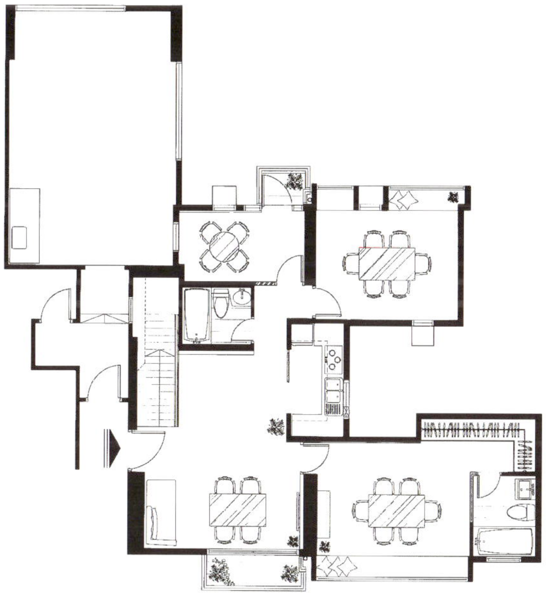

To generate a sens
fish pond undernea
installed at the fo
at the sides.

Residence A at Riverain Valley

The designer has awarded this double-level house with a modern and simple oriental beauty. It is a very challenging project as the designer has to maximize the usage rate under restriction that no change is made to the original layout.

To generate a sense of relaxation, a fish pond underneath the staircase is installed at the foyer with candles at the sides. In the living room, the simple wall is illustrated by white wooden stripes and clear mirror to facilitate visual expansion.

A gradation effect is created with the installation of a raised plane bridging the living and dining rooms. An arched walnut door is placed in the center along the wall and ceiling; and supported by a partial, translucent screen accentuated by vertical glass stripes, a touch of simple oriental aestheticism prevails. Vertical mirror stripes are applied on the wall to bring reflections of the theme wall in the living room where simplistic wooden chairs with white cushions are furnished to match the dark wood dining table set. Located on the top floor are the master bedroom, guest room and tea room which are all styled in contemporary leisure.

Surrounded by high-rise buildings with no appealing views, a canvas screen house is placed in the garden; and in order to deliver suburban sensations, tall greeneries are planted along the exterior of the property.

e of relaxation, a

th the staircase is

oyer with candles

御骏居（1）

设计师以摩登、简约的东方美来演绎这幢两层高的洋房示范单位。她回想设计这个住宅的时候，空间本不太方正，如何在不改间隔的情况下变得实用，正考验设计师的功力。

为了营造悠闲的感觉，她特别在玄关处、一道楼梯底下，做了一个鱼池，两旁放一些蜡烛作点缀。来到厅堂，她率先为客厅打造一幅别具特色的墙面，用上白木条和清镜组成一幅图案简单的墙身，清镜的铺贴使空间看来更开阔。

客厅和餐厅并不设于同一个平面，中间有一级地台的落差，这增加了空间上的层次感。在两区中央，设计师沿墙身和天花做了一道胡桃木拱门，再用上数块直条清玻璃做一道局部的、通透的屏风，立刻营造了简约的东方美。她继续以直条子的清镜来铺贴餐厅的墙身，映衬对面的、在客厅里的主题墙。她也特别为客厅和餐厅配衬线条简洁的家具，例如在客厅里的一套木座椅连白垫子、在餐厅的一组深木色餐桌椅。楼上一层分别是主卧室、客房及茶室的设计，都以摩登悠闲为主要风格。

从客厅望出去，外面布满高楼，景观一般，设计师于是在花园的位置设了一个帐布小屋，又在户外种植了很多高身植物，营造了一个在都市中的园林景致。

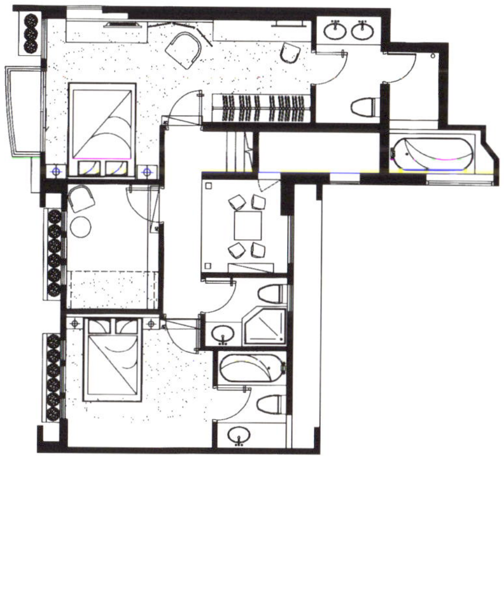

Residence B at Riverain Valley

In the limited space of show flat in Happy Valley, red and white are adopted as theme colors for the promotion of energized and vibrant lifestyles to attract young couples and single buyers. Wood flooring enriches the warm and cozy homey touch while the theme wall of bold horizontal white marble stripes and red glass complements triggers trendy sophistication.

With the limitation of space, furniture selected is light and slim. A contemporary pendent light is suspended above the round glass top dining table with white coordinating chairs. A metallic curtain, which defines the dining area and study room, is the perfect illustration of contemporary lifestyle in this compact home. The glass cabinet not only displays the owner's favorite collections, but also serves as a decoration to the wall. As there is one large window in the sitting room, natural sunlight is captured into the home.

The most conventional treatment to broaden spatial dimension is to dismantle a hard wall, however, in order to display the original layout, the designer exploited the corner of the dining area to include a study room so that the functionality of the home is enhanced. Also colored in red and white; the bedroom, which highlights a suspended bed and light trough, echoes the living room design in energized vibrancy.

御骏居（2）

为了美化这个空间不大的示范单位，设计师遂以红白作调子，营造了充满活力、轻快的生活质感，来吸引单身贵族或夫妻二人的住户。木地板的铺设加强了家的温暖感；白云石配红玻璃的主题墙身以粗横条设计，也营造了时尚味道。

由于单位的空间所限，家具的拣选也是以薄身和轻巧为主。一盏摩登的吊灯垂挂在简洁的玻璃圆桌配白餐椅组合之上，再配衬一道金属帘子轻松地隔开了书房和餐厅区域，是设计师在这有限的空间内，为时尚主题作出的最完美的演绎。玻璃饰柜的设计，既可予户主摆放心爱的装饰品，又美化了墙身。客厅的落地玻璃窗连小阳台的设计，把户外的阳光大大地纳入家中。

一般加强空间感的做法，多是以拆走实心墙为手段，然而设计师为展示空间的间隔，她采取了偷位的方法，在餐厅一角灵活地变出了一个书房空间，使这个家再增添一个空间上的功能，又与以红白为调子的睡房与厅堂的设计互相呼应。在主人房内，悬浮的睡床以及灯槽的设计为寝室空间增添了一种活泼的气氛。

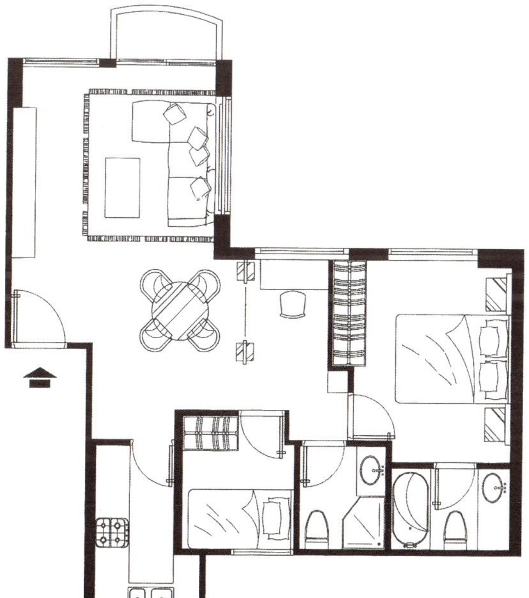

Like an urban pa
is a unique fusion
modern accentuat

Royal Green

Like an urban palace, the manor is a unique fusion of classic and modern accentuations. Characterized by golden champagne color in the entire living hall complemented with crystal curtains, pearly mosaic tiles and clear mirrors, the home displays extravagant decorative ornaments at different corners. While illumination penetrates through the light troughs and crystal chandeliers, reflections from the decorative ceiling mirror promote the grandeur of a palatial dwelling.

In order to avoid an elongated perception, a clear mirror is placed on the wall facing the French windows to broaden the spatial breadth. In front of the lavish living room is a classic dining area annexed with a contemporary open kitchen. Though both areas are presented in different styles, they are harmonized by the same color tone. Separated to the dining room through a glass wall and crystal curtain is a study room which allows visual penetration to the living area making the person at work still feels like being embraced by the glamour of a golden palace.

Settled on the top floor are the bedrooms for this regal home featuring magnificent designs even at the guest room. The highlight undoubtedly rests on the palatial master bedroom presenting a combined bedroom and walk-in closet. The crystal curtain behind the headboard conceals an imperial style closet door; complemented with a crystal chandelier and regally flaired classical fabric embellishments, it promotes a prevailing flavor of a royal bedroom. Facing the master bedroom is an aerial garden characterized by palatial landscaping where a vintage round bath tub sits on an outdoor plane. Enjoying the luxury of monarchical extravagance under the glittering stars in a regal garden, it delivers superb sensations beyond linguistic expressions.

lace, the manor

of classic and

ions

御皇庭

设计师锐意将这里打造为一个结合现代与古典风格的都市中的宫殿。走到大厅，会发现整个空间被设定成香槟金色的调子；水晶帘子、珍珠色马赛克瓷砖与清镜是装饰上的主要素材。室内不同角落也有华丽的装饰摆设，加上天花装饰镜，灯光自灯槽和水晶灯渗出，满室散发出一种如同皇宫大宅的空间气派。

走进这个呈长方形的示范单位，为免空间予人过于狭长之感，设计师遂在对着落地窗的墙身铺设清镜，用以拉阔空间。到达金碧辉煌的客厅，前方是富古典味的餐厅，与之相连是具现代感的开放式厨房，两区风格不同，但色调上却一致。在布局上，与餐厅只有一面玻璃幕墙与一道水晶帘子相隔的书房，即使在工作中，视线亦可穿透至厅堂，依然置身在金光灿烂的宫殿中。

顶层是这个皇宫的寝室，而客房也是采用金碧辉煌式的设计。当中的重点当然落在宫廷式的主人睡房之上。为了充分利用空间，设计师将衣帽间与主人房合二为一，在睡床头、水晶帘子背后展现了极富宫廷特色的衣帽柜门。睡房中央设置一盏水晶吊灯，配合古典皇宫式的布艺，令人犹如置身在皇宫的卧室里。睡房外是空中花园，设计师将之塑造成宫廷式的园艺，圆形浴盆立于户外地台之上。繁星点点加上宫廷园景，营造出仿如帝王般的生活环境。

s are Indian, the

e is infused with

f cultural preference

京士柏山（2）

简洁素雅是这个家予人的整体观感。据设计师说，因为屋主一家是印度人，这里的装潢是以他们的生活模式、文化喜好来设计的。

在入门处，玄关的整幅地面以蓝绿调子的云石马赛克拼砌出一棵细致优雅的菩提树图画，既为这里带来气派，也切合了印度的宗教文化，屋主也很喜欢。步入客厅，地面的中心部分以白色云石铺设，四边用黑色云石作框架，营造了对比的色彩和高贵典雅的气质。设计师在电视墙上，左右两旁以古典方形图案自天花至地面拼出两条宽直条，灯槽与聚光灯的配合打造了一种舒适和谐的气氛。与之相对是沙发背的墙面，墙上挂了一幅巨型壁画，两旁均设计了用古典方形图案饰边的拱门，门廊下是一些雕塑摆设。

走到二楼，右方是餐厅与休闲厅，厨房与餐厅相连。餐厅的设计呼应了客厅，色调也是以淡雅为主，墙身也饰以一个古典方形图案饰边的巨形框架，工人房连浴室则立于厨房旁边。楼梯的左方分别是两间客房和客用浴室。

顶楼设有宽阔的主卧室连浴室、衣帽间，以及两间子女的套房。设计师说，主卧室的设计，为突出女主人的温柔性情，特别用了质感柔软的帘子作为床背墙身的设计，来营造温馨柔和的感觉。

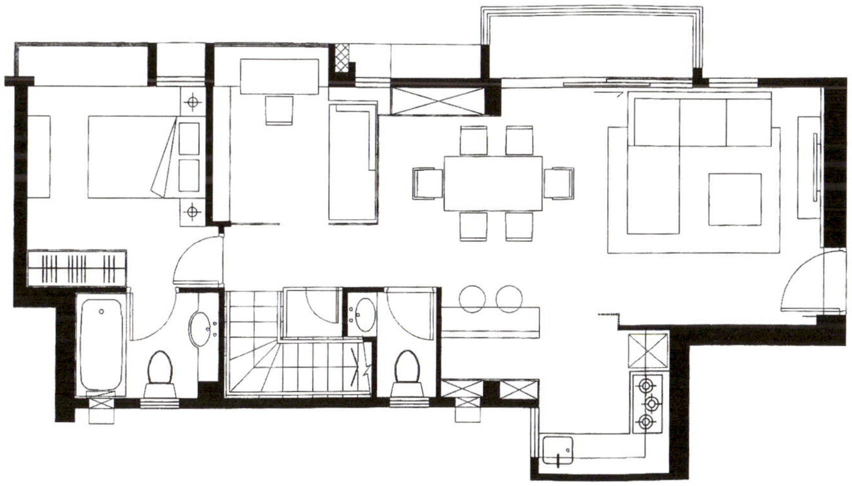

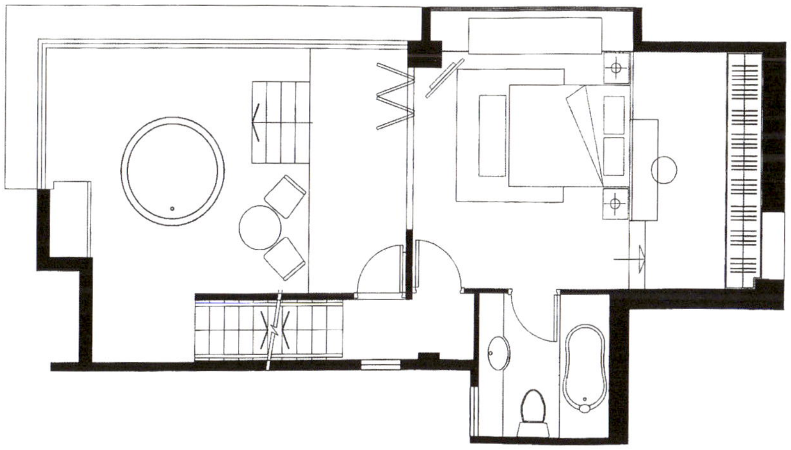

As the homeowners
design of the house
their lifestyle and

Residence B at Kings Park Hill

This home gives an impression of elegant simplicity. As the homeowners are Indian, the design of the house is infused with their lifestyle and cultural preference.

The entire floor at the foyer is covered by bluish green marble mosaics displaying an intricate bodhisattva which not only promotes grandeur extravagance, but also matches the Indian religious culture. In the living room, the central floor is covered by white marble with black marble borders to boost bold color contrast and elegant modishness. Classical squares which extend from the ceiling to the floor produce two wide stripes on the sides of the TV wall to complement with the light trough and spot lights for a comfortable and harmonious ambience. At the back of the sofa is a big mural accentuated by an arched door in classical square design where sculptural artifacts are displayed.

On the right of the second floor is a family room and dining room which connects with the kitchen. Echoing the living room design, light color palette is applied in the dining room walled by decorative frames in classical squares. At the side of the kitchen are the maid's room and bathroom while on the left of the staircase features two guest rooms and a guest bathroom.

The top floor houses a master bedroom with full bath and closet, and two children's suites. To dramatize the femininity of the wife, the designer has handpicked blinds with soft materials to be placed at the back wall of the bed for a touch of tenderness warmth.

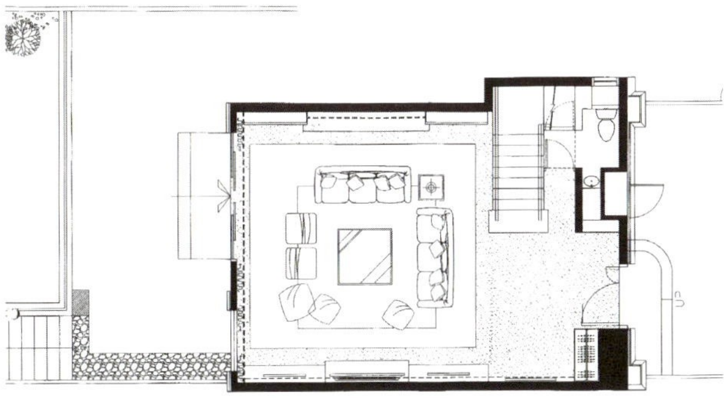

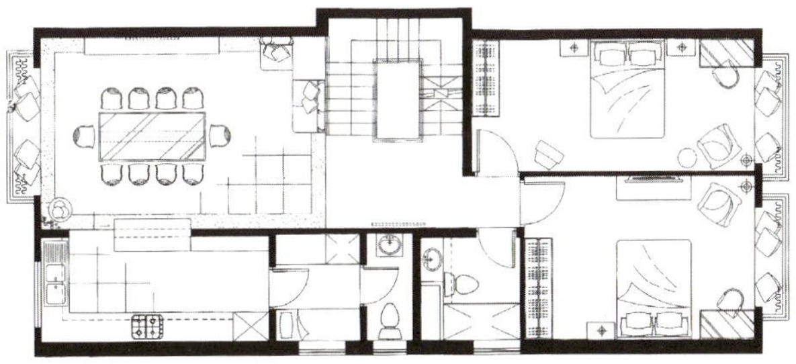

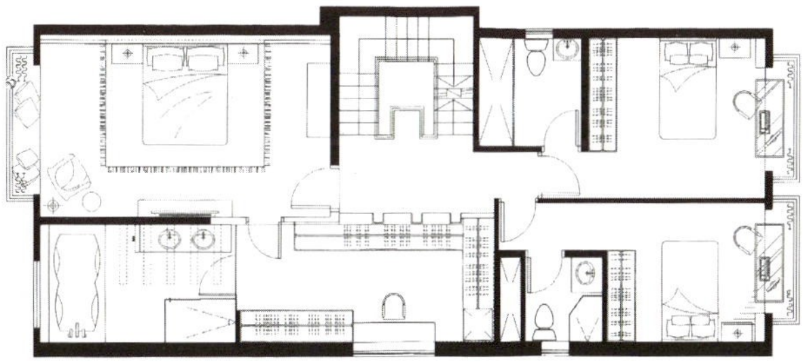

to generate a tr
metropolitan resi
has selected blac
metallic silver as

Shining Heights

In order to generate a trendy and chic metropolitan residence, the designer has selected black and white and metallic silver as the theme colors. The bedroom, however, is characterized by brown, red, and green for a statement of cheerfulness to break the norm.

This show flat is a four bedrooms apartment. To create a generously spacious master bedroom, the designer has dismantled part of the walls of the closet and study room; and to take advantage of the exceptionally high ceiling which exceeds 3 m, two-step and four-step platforms are built in the study room and closet. A feeling of tiered ascendance can be clearly experienced when people penetrate the bedroom through the study room and closet. As the bedroom floor is raised, it levels with the window sill turning the window into a French window to enjoy the fantastic outdoor scenery.

Applying a unique design that emulates a recessed bath tub, an elegant master bathroom is created. The designer opined that it will be very costly to actualize the design through structural means. However, with creative refurbishment, an outstanding effect can also be achieved.

For zoning definitions and contrasts, the ceiling heights in the living and dining rooms are different. Distinctive screen panels are also applied to further define the various zones. The foyer is characterized by a striped glass panel while the living and dining rooms are highlighted by either a framed mirror round screen or a thick silver frame screen. White in color to maintain a coherent color scheme, the sofa sets carry unique styles to promote a lively and energetic visual effect.

A wine cabinet and glass wall are installed in the kitchen to deliver a translucent openness; at the same time, the glass blocks cooking smokes from emitting to the living room area.

Colored in brown, the master bedroom may not seem to be consistent with the main color scheme. The designer's intention is to promote sense of liveliness, like the use of red and light wood in the daughter's bedroom and green in the son's bedroom. The room furniture as seen in the wardrobes and student desks, however, come from the same series. Therefore, even though the colors differ, they carry an appealing consistency that echo the main theme.

dy and chic

dence, the designer

k and white and

the theme colors

亮贤居

为打造一个时尚耀眼的都会居所，设计师在整体空间上选择以银、白、黑作为主色调；然而，她在睡寝区刻意用上咖啡色、红色与绿色，来带出活泼的感觉，突破惯常使用的黑白灰。

这个住宅示范单位拥有四房两厅，在空间规划上，设计师把分隔了主卧室、衣帽间和书房的部分墙身拆去，制造了一个宽敞无限的主人套房。由于这个单位拥有极高楼层，更高逾3米多，设计师将这个优势打造了一个特别的睡房设计：她分别在书房与衣帽间加建两级及四级地台，当人们从书房走到与衣帽间相连的寝室时，会产生一种在空间上层层递进的感觉。由于地台被升高，睡房的地面与窗台继而被拉成同一个高度，使立于窗台上的一扇窗成了一面可尽览窗外美景的落地玻璃窗。

在主浴室，设计师利用巧妙的方法把浴缸做到仿如下陷在地里，营造一个高贵的浸浴空间。她说若在结构工程上来体现这个设计，费用一定高昂。不过，这却可利用装潢达致极佳的效果。

回到客厅与餐厅的天花，高度上并不划一，为的是突显分区，带出一个对比。设计师也会采用不同的物料做屏障来划出区域，例如用条子玻璃屏风点示出来的玄关区或以圆形框子覆盖的一幅镜面、粗银色镜框等装饰来点示厅区。为配合整体空间的色调，客厅的沙发家具全选了白色，不过却是不同的款式，增加了活泼跳脱的视觉效果。

至于厨房的设计，设计师以酒柜及玻璃作墙身来划出厨房的空间；使用玻璃作墙身的好处，一方面予人开放式的通透感，同时又可避免油烟散出厅区。

设计师想用多种颜色带出活泼的感觉，如女孩的睡房是以红色和浅色木作主调，男孩的则是绿色为主；尽管深咖啡色的主卧室在色彩上与厅区并不连贯，当中的家具在风格上却是一致的，例如衣帽间内的衣橱与书房内的书桌是出自同一个系列的。所以，纵使色彩不一，别人却不觉整体的设计是杂乱无章，而是呼应主题，带出一种美感。

Sexy Girl

PUTMAN
STYLE

DONNAKARAN
NEW YORK
DONNAKARAN

CHATEAU BIRÉ
BORDEAUX SUPÉRIEUR

ADVENTURE
ADVENTURE

NEW YORK
detail

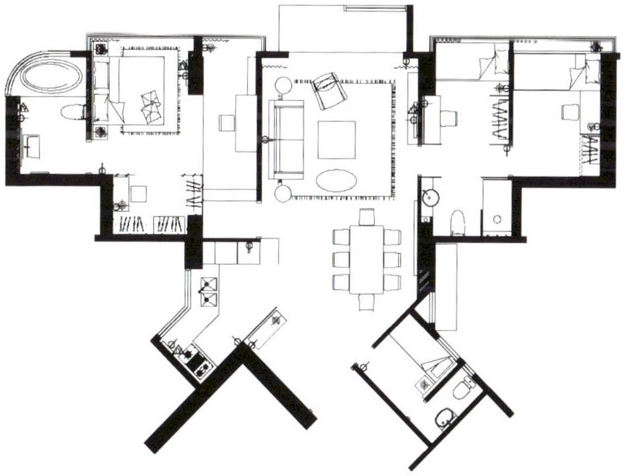

thick palm forests
wonderful rainfore
which inspired the
palatial thai

Casa Marina

The designer has been fully attracted by its outdoor environment when she made her first visit to this residence. Green mountains and thick palm forests prompted her to the wonderful rainforest of Thailand which inspired the adoption of palatial Thai approach as the design theme.

Possessing a rectangular layout, the entrance is located at the side of the living room. While the dining room is on a raised platform and connected to the living room through a flight of steps, the kitchen joins the dining area and links to the bathroom, laundry room, and maid's room. The second floor houses two guest suites and a family room while the master suite and study room are located on the top floor.

Although themed in Thai predominance with generous application of wood and silk, the designer has also adorned the house with Indonesian sentiments as characterized by the sandstone bricks and wooden sculptures in the dining area. The theme wall in the living room displays a country-style culture stone while the opposite wall is mounted with an antique arched door dated over a century. It was originally the door of an old church but now has become the focal point of the residence. An antique screen panel is also erected in the master bedroom which goes along with the owner's unique taste, and accentuates the home with an opulent allure of history. The ceiling of the living room features soft suspended draperies for a warm and romantic appeal while the dining room adopts the traditional Thai conical ceiling and rooftop design. Both areas are furnished with exclusive Southeast Asian pendent lights to trigger relaxing sentiments of passion.

The furniture and fixtures all deliver basic and minimalist aestheticism. The wooden chair with its matching white cushion in the living room, the dark wood round dining table and chairs in dining room, the dark wood furniture in the bedroom… all coordinate impeccably with the home décor to promote an ambience of serenity. The master bedroom is the designer's favorite; in order to create a sense of royal supremacy, a two-step raised platform is built for the bed. Gently wrapped by soft canopies, it accentuates caressing comfort when touched by tender breezes which also suggests for vacation indulgence.

prompted her to t

st of Thailand

adoption of

浅月湾

初次看到这所洋房的户外环境——被翠绿的山岭和棕榈树林重重包围，令设计师有如置身于富饶多雨、翠绿如染、溪流淙淙的泰国丛林里。这样的景致激发了她的创作灵感，并决定以泰式宫廷作为这个住宅的设计主题。

这个住宅的空间呈长方形，入口设于客厅内侧；客厅和餐厅并不是设于同一个平面，餐厅设在一个地台之上，两者有一个落差，中间以一道楼梯作过渡。厨房设于餐厅旁边，与卫浴室、洗衣房、工人休息室相连。二楼分别设有两个客房连浴室，以及一个家庭厅。顶层是主人房连浴室、书房的设计。

虽然泰式风格是这里的主调：木材和丝绸物料的运用随处可见，设计师也揉进了点点印尼风情，如铺设在餐厅的砂石砖和一些木雕摆设。走进客厅，主题墙上采用了带有户外田园风味的文化石，与之相对的墙壁，则镶嵌了一扇超过百年历史的古董拱门。据设计师说，它原来自一幢旧教堂的门口，在这里顿成了整个空间的焦点。她还把一扇怀旧屏风放置在主卧室内，这些古董收藏实在非常切合户主的独特品位，也为这个家带来浓厚的历史感。回到厅堂的装潢，设计师在客厅的天花，以帐幔的柔和质感缔造温馨、浪漫的感觉；餐厅的天花位置则呈尖锥状，这也是仿照传统的泰式屋顶而设计。两个区域分别悬挂了充满东南亚色彩的别致吊灯，灯光投影于天花上，营造了随心的浪漫。

在家具配置上，简洁朴素是其主要风格，例如放置在客厅的木座椅配白垫子、餐厅里的深木色圆桌与餐椅、睡房里的深木色家具……这些木制家具与一室的装潢布置非常合衬，更带出一片和谐静谧的气氛。设计师尤喜欢主卧室的设计，为了营造皇室尊贵的气派，她在睡床下加建两级地台，并于床上挂上轻柔的布幔，当风轻轻吹过，予人悠闲舒适之感，这里也弥漫着一种度假风情。

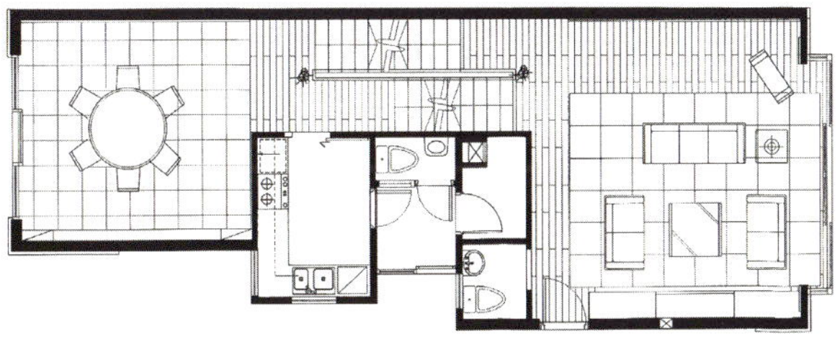

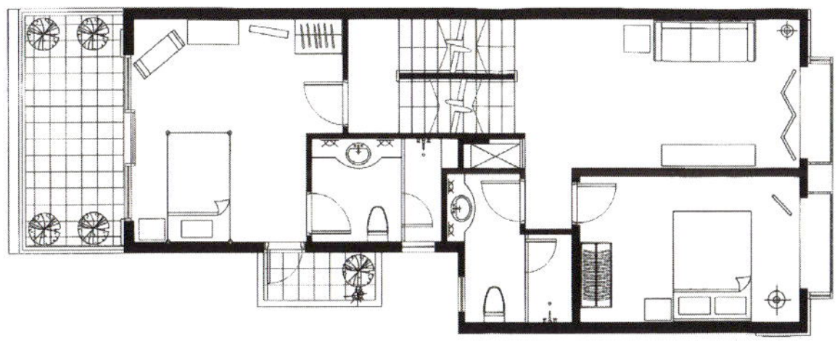

This home is infil[...]
unique characterist[...]
Art Deco.

Residence B at MacDonnell Road

The Art Deco approach of the early 20th century is the designer's favorite for its revelation of quality lifestyle; its pursuit for functionality, color and structure has triggered many classical designs. In this 600 sq. m show flat located at MacDonnell Road, the designer fully demonstrates the details and aesthetic elements of Art Deco.

Developed after WWI, Art Deco is a design movement to counter balance the mundane of mass production and signifies the stride from traditional conservatives to contemporary sophistication. For people of today, of course we witness even more avant-garde complexities; however Art Deco does symbolize the concurrent interpretation of art and has tremendous impact to the design development of the 20th century. Even today, Art Deco is still actively presented in our daily lives in diversified forms.

This home is infiltrated with unique characteristics and details of Art Deco. At the foyer, for example, the light box composed of clear mirror and engraved glass, and the twin light panels that extend from the ceiling to the floor of the master bedroom are artistic gallery display of the era. The TV wall in the living room also features a classical wooden engraving which echoes the arched door design in the living room complemented with clear mirrors.

The design of furniture and fixtures also features refined decorative details: the leather chaise lounge next to the sofa in the living room is a concurrent design of the Art Deco era; the coffee table near the chaise lounge and the table legs in the dining room are all replicates of that time.

The house carries prevailing moods of elegance and modernity. Floored in cool marble tiles at the living room, it contrasts boldly with the warm wood flooring in the dining area. The ceiling of both areas features step like design which enhances an appealing layer effect. Since the working desk is installed at the balcony, it provides a refreshing sea view for relaxation. In order to avoid monotony, one of walls in the master bedroom is covered by dark wood panels complemented with stainless steel fittings. A tiny and dainty display cabinet is furnished at the corner of the closet to make it appear more compact and light weight.

trated with

tics and details of

麦当奴道住宅（2）

设计师一直喜欢20世纪初装饰艺术（Art Deco）的风格，认为它具体呈现一种生活的态度，坚持对色彩、形式与实用性的追求，从而创造了很多经典的设计。她在这个位于麦当奴道、占地600平方米的示范单位里，也尽情地注入了这种风格的美学元素与细节。

Art Deco是第一次世界大战后兴起的一种美术运动，抗衡了大量机械化生产的复制品，也代表了由传统风格走向摩登的特色。当然，对现今的人来说，比 Art Deco 更新更摩登的风格也有很多，然而Art Deco也象征了一个时代的艺术表现，其影响力更远及20世纪设计史的发展，至今依然灵活、多变地出现在我们的生活当中。

这种风格的特色和细节，在这个家里俯拾皆是：例如在入门处，设于玄关位置的一具以清镜和雕花玻璃做成的灯箱。通过进口走道直达主卧室，就在睡床的两旁，分别镶嵌了两幅自天花延伸至地面的灯板，都是那个时代用来设计门廊的艺术图案。回到客厅，电视墙身也使用了一幅经典的立体木雕装饰作设计，这设计亦延续至餐厅的一幅墙上、以一个拱门的形态铺贴其上，中间部分缀以清镜。

设计师在家具设计上，也渗进了一些装饰细节：在客厅里、沙发旁的一张高雅的真皮贵妃躺椅，也参照了Art Deco时代的家具款式设计；躺椅旁的茶几以及在餐厅里的餐桌脚，也仿照了当时的家具形态来塑造。

环顾这个家，予人的感觉是既典雅又摩登的。客厅的地面以云石铺设，与餐厅的木质地板，一冷一暖形成了对比。两个厅区的天花均采用了叠级的方法来处理，增加了层次感。走到书房，书桌设于窗台位置，面向令人心旷神怡的海景。主卧室的设计，以深色木铺缀一面墙身，为了不过于单调，加入了少量的不锈钢元素作饰边。在衣柜的旁边，其弯角处是一个小巧的陈列装饰架，这使柜身看来更轻巧。

ANDRE
TORRES

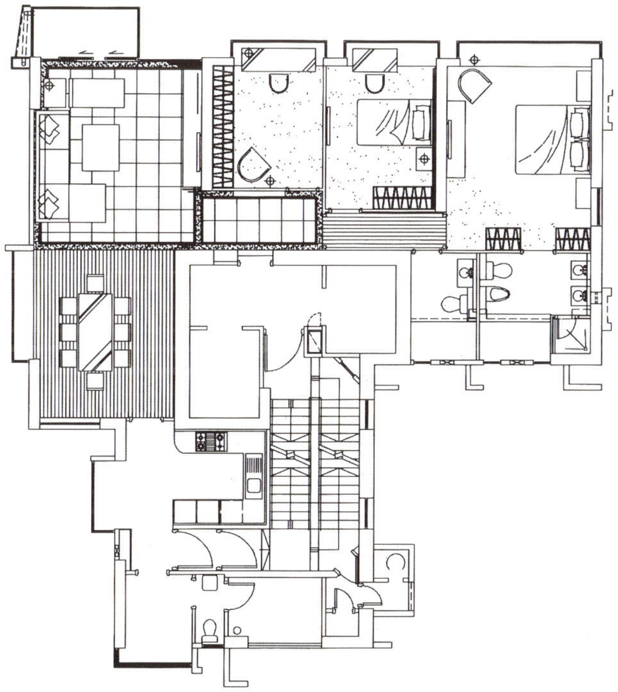

Residence B at Grand Waterfront

This show flat is characterized as a glittering star that sparkles on the harbor to symbolize Hong Kong's diversity, modernity and contemporary charisma.

The design concept was developed at a time when silver color was hot in fashion. Inspired by its vibrancy and glamour, the designer decided to use the color to express contemporary classics and applied it from floor to walls. Featuring curvilinearity throughout the proposition, the central entrance displays symmetrical eclipsed walls on both sides which resemble a pair of arms reaching out to embrace incoming guests. The curvilinear spaces on the sides are designated as living room and dining area respectively. While the mirrored ceiling accentuates the ceiling height, the curvilinear sofa and dining chairs express minimalist aestheticism.

The living hall is the designer's favorite since it best demonstrates the lifestyle of chic and contemporary trendiness. She deliberately installed an embedded cabinet to display the 12 Zodiacal crystal figurines from Baccarat to promote a crystalline brilliance which echoes with the spot-lit glittering wall.

To maximize the benefits of sea views delivered by the French windows, a clear mirror is erected at the opposite wall to capture and strengthen its impact. Frosted glass is generously used for the wardrobe to enhance the contemporary touch. Besides the glittering silvery bedroom set, an elegant bronze curtain is suspended behind the headboard. The master bathroom is walled by glass panels to dramatize translucency and a carefree sensation.

Branded designer furniture is handpicked for various spots to dramatize the statement of grandness. Colored in pink, the girl's bedroom is themed on Cinderella while the boy's Ferrari theme takes on white. Even though the colors differ, they both possess refreshing taste of glamour. A game room is situated between the children's bedrooms for them to play and leisure reading.

翔龙湾（2）

设计师把这个示范单位塑造成一颗悬挂在海港上空的耀眼新星，并以此象征香港的多元性、现代化，以及时尚的气质。

在构思这个项目的时候，银色在时装界正大行其道。银色的耀眼夺目，吸引设计师以这种颜色来表达时尚的概念。这个单位内无论地材、墙身等，都是以银色为主。在设计特色上，她以弧形线条勾画整个空间。由于大门入口刚好在中间，于是她把左右两旁的墙身对称地塑造成弧形，就好像一双手拥抱着进屋的人客，予人一种温暖的感觉。两边的弧形空间分别是餐厅和客厅；天花上的清镜铺贴，增加了楼层的高度。在家具配置方面，简洁的沙发和餐椅等都是弧形线条为主的。

环顾不同区域，设计师最喜欢厅区，因为它的视觉效果最好，让人深深感受到时尚潮流的生活感。她也特别在厅区设计了一个入墙装饰柜，放了Baccarat12生肖的水晶摆设，以增添晶莹剔透的美，也回应着在灯光折射下闪闪生光的墙面。

从客厅来到主卧室，为了放大落地窗可看到海景的优势，设计师在房间的另一幅墙上铺贴清镜来把海景纳入室内，极大地加强了临海的感受。她也在衣柜门上用了大量磨砂玻璃镜，来加强时代感。房内亦使用了银光闪闪的床铺，床头后是一扇高雅的铜帘。主浴室的墙身被改装成玻璃，突显了这里通透无拘束的感觉。

这个空间的不同角落，设计师均选取了一些由名师设计而成的家具作摆设，来突出奢华的感觉。两间为小孩而设的睡房，粉红色作主色的是为小女孩设计，主题是灰姑娘；白色的是给男孩子，主题是法拉利，两者空间的色调不同但同样具闪耀却清新的效果。位处两间孩子房的中间，设有一间游戏室，可供他们阅读或玩乐之用。

Miele

My Secret Diary

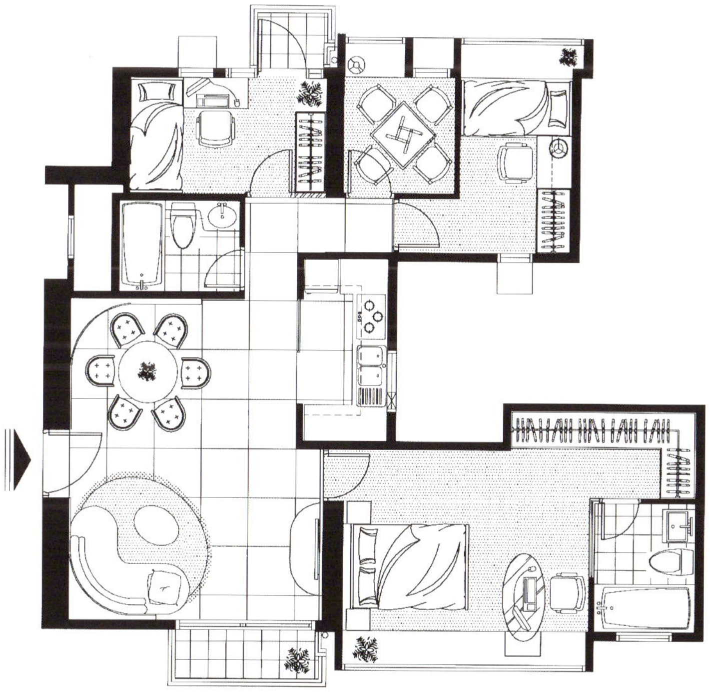

Acknowledgements 鸣谢

You are the April spring of earth

Famous Chinese architect Lin Huiyin has dedicated the poem of love -'I said you are the April spring of earth' to her son; now I would like to bestow the same statement to my beloved mother. Without her nourishment and encouragement towards my artistic talents and persistent support on my educational endeavours, I could never become an autonomous interior designer. It is a phenomenal vision during that time in Hong Kong for a mother to possess the wisdom in recognizing the importance of education and to inspire her children to move forward. Without her dedicated love, this book could never come to being. Mother, I truly thank you.

I would also like to offer my special thanks to Mr. Alex Lam for his trust and recognition. He is not just a good friend to me, but a catalyst who ignited my initial contacts with show flat design projects which opened up opportunities for my development and further expansion. Being my first residential and show flat design collection, this book could never be published without the support and encouragement of Mr. Lam. For this, I thank you.

The love and care from parents and friends is like the gentle sunshine in April spring which warms up my heart. I would like to share this tender radiance to every single one of my readers.

'You are the blossom of every living tree;
You are the swallow that whispers between ceiling beams;
You are passion, warmth, hope,
And the April spring of earth.'

Anita Tang
Jan 2010

你是人间的四月天

“我说你是人间的四月天”。中国著名女建筑师林徽音把这首爱的颂歌献给儿子，现在我想借这句话送给我挚爱的妈妈。如果她没有发掘我的艺术天份并在教育上给予的栽培、鼓励与支持的话，我绝不可能成为一个独当一面的室内设计师。在那个年代的香港，一位母亲能如此独具慧眼，明白学习的重要，是一件非凡的事，也能启发孩子走得更远。如果没有她的爱，这本书根本

在这里，我也要感谢一位朋友，林高禹先生，我感谢他对我的信任与肯定。林先生既是我的好友，也是我的伯乐。还记得当年就是他给我机会，让我开始接触样板房的设计工作，开展这方面的设计领域。这本书是我第一本家居设计专刊，如当初没有林先生的支持和鼓励，这本书也是难以出版的。林先生，我也感谢你。

父母与朋友的关怀，有如四月天春和日丽的阳光，温暖我的心怀，我在此谨把这份温暖，与各位读者共享。

“你是一树一树的花开，
是燕，在梁间呢喃，
— 你是爱，是暖，是希望，你是人间的四月天。”

邓凤贤，2010年1月

图书在版编目(CIP)数据

写给空间的情书 / 贝思出版有限公司汇编.—沈阳：辽宁科学技术出版社，2010.1
ISBN 978-7-5381-6221-9

I. 写… II. 贝… III. 住宅—室内设计；空间设计—图集
IV. TU241-64

中国版本图书馆CIP数据核字(2009)第229786号

出版发行：辽宁科学技术出版社
(地址：沈阳市和平区十一纬路29号 邮编：110003)
印 刷 者：广州泽美印业有限公司
经 销 者：各地新华书店
幅面尺寸：283mmx292mm
印 张：17
插 页：4
字 数：50千字
出版时间：2010年1月第1版
印刷时间：2010年1月第1次印刷
特邀编辑：吴嘉慧
责任编辑：郭 健
封面设计：梁佩琼
版式设计：梁佩琼
责任校对：李 霞

书 号：ISBN 978-7-5381-6221-9
定 价：248.00元

联系电话：024-23284536
邮购热线：024-23284502
E-mail:rainbow_editor@163.com
http://www.lnkj.com.cn
本书网址：www.lnkj.cn/uri.sh/6221